Project Genesis

Project Genesis

✦

Decoding the Universe

Ian Beardsley

iUniverse, Inc.

New York Lincoln Shanghai

Project Genesis
Decoding the Universe

iUniverse books may be ordered through booksellers or by contacting:

iUniverse
2021 Pine Lake Road, Suite 100
Lincoln, NE 68512
www.iuniverse.com
1-800-Authors (1-800-288-4677)

ISBN: 0-595-34490-9

Printed in the United States of America

Contents

Since other stellar systems may not even exist as we need them, and the distances between them are so immense, it might be better to unlock the mysteries of making them, and find the structure in ours that allows for so much life. There is, I have found, a correlation between the microworld and the macroworld, where our solar system is concerned. It may be related to why it is life bearing.

Part 1

An interesting family of substances is methane (CH_4), ammonia (NH_3) and water vapor (H_2O). Methane is tetrahedral in structure, a carbon atom sourounded by 4 hydrogens. Ammonia is trigonal pyramidal, a nitrogen atom surrounded by 3 hydrogen atoms, and water vapor is triangular, or bent, an oxygen atom surrounded by two hydrogens. These represent stable structural systems as they are all systems of triangles, which are the only stable polygons. These substances combined under energy with hydrogen gas form amino acids, the building blocks of life. The core atoms of these molecules, carbon, nitrogen, and oxygen, are all in period two of the periodic table and follow directly one after the other, and are all in amino acids, the hydrogen as well. It is a hypothesis of astrobiology that amino acids formed in the protoplanetary cloud before the earth ever formed. In this sense we may have our origins in deep space. Is what I mean by structural systems is that there are only three structural systems, the tetrahedron, the octahedron, and the icosohedron. They are the only stable solids, that is noncollapsing flex corners whose faces are triangles. Most compounds are something other than these, like pentagons with linear off shoots for example, that comprise the wrong number of atoms to make a "solid" unit, and I mean solid as in the pythagorean solids, the geometric term. Both methane and ammonia make different variations on the tetrahedron, a pythagorean solid.

When plants perform photosynthesis, they combine carbon dioxide with water and release oxygen. The reaction is:

CO_2+2H_2O-‡CH_2O+O_2+H_2O

As can be seen a sugar is made. Important to most plants to do this is Nitrogen. Nitrogen (N_2) is the most abundant gas in the earth atmosphere, comprising about 78.03% of it. We now calculate the molecular masses of these special gases:

CH_4=(12.01+4(1.01))=16.05
NH_3=(14.01+3(1.01))=17.04
CO_2=(12.01+2(16.00))=44.01
H_2O=(2(1.01)+16.00)=18.02
N_2=(14.01+14.01)=28.02

$O_2=(16.00+16.00)=32.00$

We now form some ratios between these molecular masses:

$(O_2)/(CH_4)=32.00/16.05=1.992\sim2$

$(NH_3)/(CH_4)=17.04/16.06=1.061\sim1$

$(CO_2)/(O_2)=44.01/32.00\sim1.4=sqrt(2)$

$(CO_2)/(N_2)=44.01/28.02\sim1.6=(sqrt(5)+1)/2=phi$

$(O_2)/(H_2O)=32.00/18.02=1.776\sim sqrt(3)$

Notice that these values are given by the sequence:

$|2cos(pi/n)|$ $n=(1,2,3,4,5,6)(pi/n)$radians

Observe:

$2=|2cos(pi)|$

$0=|2cos(pi/2)|$

$1=|2cos(pi/3)|$

$sqrt(2)=|2cos(pi/4)|$

$(sqrt(5)+1)/2=phi=|2cos(pi/5)|$

$sqrt(3)=|2cos(pi/6)|$

Geometrically sqrt(2) is the ratio of the side of a square to its radius. Phi is the ratio of the chord of a regular pentagon to its side. Sqrt(3) is the ratio of the side of an equilateral triangle to its radius, and 1 is the ratio of the side of a regular hexagon to its radius. The square, the regular hexagon and the equilateral triangle are the tessellating regular polygons. The regular pentagon is one of the archemedian tessellators.

Part 2

We compare the mass of the earth to the mass of the sun, and multiply that ratio by the distance between them. Let the mass of the earth be M_e, and the mass of the sun be M_s. Let the distance between them be r.

$(M_e/M_s)r=(5.976E27/1.989E33)(1.495979E13)=(4.495E7)cm=449.5km$
We now divide that result by the radius of the earth, R_e:

$(449.5km)/(6378.5km)=0.07$

Hydrogen is the most abundant element in the universe and Nitrogen is the most abundant element in the earth atmosphere. We now compare their molar masses:

$(H/N)=(1.01)/(14.01)=0.07$

And we see that

$(H/N)=((M_e)(r))/((M_s)(R_e))$

Having showed the last equation, where hydrogen is the most abundant element in the universe and nitrogen is the most abundant element in the earth atmosphere, then since Mars is a terrestrial planet upon which we can set foot as opposed to Venus and Mercury, let’s apply the same idea to mars. The most abundant gas in the Mars atmosphere is carbon dioxide, or CO_2. It in fact comprises 95.3% of its atmosphere. We have:

$H/(CO_2)=1.01/44.01=0.02$

Now let M_m= mass of mars, M_s = mass of the sun, r= the distance between them, and R_m= the radius of mars. We have

$$(M\_m)(r)/(M\_s)(R\_m)=(6.418E26)(2.279409E13)/$$
$$(1.989E33)(3.393096E8)=0.02$$

And therefore,

$$H/(CO\_2)=(M\_m)(r)/(M\_s)(R\_m)$$

Keep in mind these equations, both for the earth and mars, hold for a solar system at its peak as an orderly arrangement of parts. Eventually it will begin to degenerate. The sun is losing mass every day and therefore r, for any of the planets, will grow.

Thus we say in general:

$$H/A=(M\_p)r/(M\_s)(R\_p)$$

where H is the molar mass of hydrogen, A is the molar mass of the most abundant element or gas in the planet's atmosphere, (M_p) is the mass of the planet, (M_s) is the mass of the star, r is the distance between the planet and the star and (R_p) is the radius of the planet. Lets look at the quantity (M_p)r/(M_s). It is equal to (d_1)/(d_2)(d_1+d_2), the ratio of the distances between the balancing point of a cosmic teeter totter and the planet and the star balanced on it, times its length. We then compare such a distance to the radius of the planet.

Part 3

The relative equatorial surface gravities uncorrected for centrifugal force of the earth and mars respectively are 1.000 and 0.380. Their proportions are

1.000/0.380=2.63

The ratio of the molar mass of oxygen gas to that of carbon is

$(O_2)/C=32.00/12.01=2.66$

Thus, $(g_e)/(g_m)\sim(O_2)/C$

where g_e is the equatorial surface gravity of the earth and g_m is the equatorial surface gravity of mars. The centrifugal forces being nominal, this says it takes the same amount of energy to lift a mole of carbon on the earth as it does to lift a mole of oxygen gas on mars the same distance if the atmospheric pressures are excluded. Carbon is the basis of life and oxygen gas its necessity (for human life).

The data for this study came from the Handbook Of Space Astronomy And Astrophysics, by Martin V. Zombeck, Cambridge University Press, 1982.

Heart of Genesis

Luminosity of the sun=3.826E26J/s=L

seconds in year=3.1536E7s=t

Mean orbital velocity of earth=29790m/s=v

Mass of the earth=5.976E24kg=m

Lt=1.2E34J

$(1/2)mv^2 = 2.65E33 =$ kinetic energy of earth

$Lt/(1/2)mv^2 = 4.5$

Now consider the molar masses of iron and carbon:

$Fe/C = 55.85/12.01 = 4.6$

Thus
$Lt/(1/2)mv^2 = Fe/C$

Thus the comparison of the annual energy output of the sun in light, to the kinetic energy of the earth, or to its energy of motion in other words, is the same as the comparison of iron to carbon as far as the weight of an atom is of the former to the latter. Keep in mind a year represents one complete revolution of the earth about the sun and that the iron age represented a revolution in tool making while carbon is the basis of life. We have used the 365 day year for this calculation, and only considered the kinetic energy of the earth do to its orbital motion. It is natural now to ask what is the comparison of the density of Iron to that of carbon at earth temperatures and pressures:

$Fe = 7.87g/cm^3$ and $C = 2.26g/cm^3$

$7.87/2.26 = 3.48 = Ag/P = 107.9/30.97$

Ag is silver and P is phosporus. Silver is the most conductive metal of heat and electricity and the most reflective, being used therefore in electronics and to make telescope mirrors. Phosphorus is necceary to animal and plant life and is returned to life after used in one of the four natural ecological cycles.

Washington Alloy Co writes:

"Washington Alloy Phos-Copper-Silver Brazing Alloys (USA 0, USA 2%, USA 5%, USA 6F and USA 15%) are all manufactured to offer economy as well as consistently high standards of quality and performance. These alloys are excellent for joining copper to copper where the phosphorus content of the phos-copper-silver brazing alloy reacts with the copper of the base metal in such a way that the filler metal becomes “self-fluxing”. For this reason these alloys are used quite extensively for joining closed copper tubing in the refrigeration and air conditioning industries where flux removal after brazing is difficult to impossible."

World Book 2004 writes:

"Copper is the best low-cost conductor of electric current. As a result, the electrical industry uses about 60 percent of the copper produced, chiefly in the form of wire. Copper wire carries most of the electric current inside homes, factories, and offices. Large amounts of copper wire are used in telephone systems, as well as in television sets, motors, and generators."

Thus the mechanics of the solar system have taken us to copper, among other things, all key to technology in their various ways. Copper occurs free in nature, but we get most of it from copper sulfides. We separate it from copper sulfide in the following reaction:

$Cu_2S(l)+O_2(g)$--->$2Cu(l)+SO_2(g)$

Thus the pollutant sulfur dioxide is released into the air, a major contributor to acid rain. If we relate Cu, SO_2 and S to Ag and P, we have come all the closer to finding the connection of human activity to the motion of the earth as related to the output of solar energy during this phase in human life where it is beginning to enter space.

We have

$(Cu_2)/S=3.96=A$ and $(SO_2)/Cu=5.33=B$

We also have that $Ag/P=3.48=C$ and

$(densityAg)/(densityP)=5.7692=D$

(densities at earth temperatures and pressures)

$AB=21.1068$ and $CD=20.07$

Thus,

$AB=CD$ within 95%

Notice that $(O_2)/(S)\sim1.00$

O_2 we breathe, but combined with S is poison.

If something is readily available in nature, odds are it is there for us to use, but the second we have to make it, extract from it from something, or alter it, we will upset the natural structure that maintains a healthy balance for life, ourselves a component of that structure. But perhaps components can change, if other components change in the right way, including ourselves (i.e. the story of evolution).

As you can see we are on a quest that is about to fullfill itself, as all of this leads us to coke and coal, I am not sayin how, but obviously it is a dead give away as its importance surmounts as we run out of oil while on the verge of entering a real space age, and that it is to energy production (electricity) and metalurgy perhaps what copper and silicon has been to electronics. It has also been used since no one knows when, but the chinese started its production as an industry in…

Thus we compare copper to silicon, and equate it to coal, C, which is carbon, compared to hydrogen and find that they are equal by a factor x, which, is the ratio of the mean jupiter-sun distance divided by the mean earth-sun distance:

$$C/H=x(Cu/Si) \quad x=5.2$$

Hydrogen is compared to coal, because it too is an energy source, the one that is proposed to take the place of fossil fuels, technological breakthrough pending to separate it from water so that more energy is provided than is put into the process. Doped silicon represented a revolution in electronics in the form of integrated circuitry. Jupiter should become our source of hydrogen gas. Hydrogen gas comprises 86% of its atmophere. Copper and coal have been man's work horse in the past, silicon and hydrogen his future. As elements of the past were important they become important again, but in a different way. Thus carbon in the form of coal, heated our homes, and smelted Iron, now it shows promise in the potential of becoming a one dimensional quantum wire. Silver and gold await their place in the future, the vast reservoirs of jewelry converted into circuitry because of their special, and important electrical properties. Silicon is on the border between the recent past and the future. Mars is the one terrestrial planet we can colonize. Thus in the spirit of our last equation we write:

$$Ag/Al=y(Si/C)$$

where Ag is silver, Al is aluminum, Si is silicon and C is carbon. "y" is the ratio of the average mars sun distance to the average earth-sun distance. y=1.52

The work horse of electronics is tin-lead solder (Sn and Pb) and the workhorse of chemistry is pyrex glass, basically silica (SiO_2) and boron (B). Let z=9.54 the ratio of the Saturn-sun distance to the earth-sun distance, and we have, in the spirit of our last two equations

$$Sn/Pb=z(SiO_2)/B$$

I think creating some categories is called for here:

Reservoirs: coal, hydrogen gas, methane etc…

Hardware: solder and solder iron, beakers, and test tubes, glassare etc…

Systems: silver wire, copper wire, aluminum wire…etc

Software: integrated circuits, transistors, diodes, resistors, capacitors, components…

We can now write by the last three equations:

Jupiter (meaning) reservoir/system/software=technology
Shelter: regular hexagon
Mars (meaning) software/system=magic
Food: square
Saturn (meaning) hardware=industry
Self: equilateral triangle

The dual of the regular hexagon is the equilateral triangle, the dual of the square is itself, and the dual of the equilateral triangle is the regular hexagon.

Invariances in nature are searched for so they can be used to form ratios that can be compared. A reason for the harmonies is sought. Intuition is utilized. Encyclpedia Galactica (math notes):The idea is to find the unifying themes of each subject, and unify those under a common theme so as to make it easier to remember and more effective in its applications.
It is becomeing increasingly clear that we should enter a space age, and, as such, only use coal and coke for metalurgy, gold for electronics and get gases for transportation from the gas giants, like Jupiter, Saturn, and Neptune (methane and hyrogen) and pipe in our electricity from states such as oregon which have a great surplus do to their immense river power. Remember, the clean burning fuel cell auto can run off of hyrogen.

The formula for the legs of right triangle given its area and hypotenuse is:

$$a,b=+/-sqrt((h^2+/-sqrt(h^4-16A^2))/2)$$

Earth surface area is 5.1E8 km^2

Solar radius is 696,000 km

Using these values in our formula, we get

a=6.957E5 km and

b=1.47E3 km

a/b=473

Now

a/b=[(mass of jupiter)O]/[(mass of earth)C]

where

mass of jupiter=318 earth masses and O is oxygen and C is carbon. We have said what O and C mean!!! This allows us to begin to decode project genesis.

O=16.00 and C=12.01

Now silicon, oddly enouph is in the same group as carbon. As we are carbon based it is interesting that artificial intelligence, or man made intelligence, would be silicon based. Silicon is pretty much useless, as I understand it, if it is not "doped" with boron or phosphorus, atleast as far as negative and positive type silicon are used. Let us consider phosophorus is for making negative type silicon and boron is for making positive type silicon. Let us find the geometric mean between phosphorus (P) and boron (B) and divide it by silicon (Si):

sqrt(P*B)/Si=sqrt(30.97*10.81)/(28.09)=0.65

and let us take the harmonic mean between phosphorus and boron and divide it by silicon:

(2*(30.97*10.81)/(30.97+10.81))/28.09 =0.57

Now let us take the arithmetic mean of these two numbers:

(0.65+0.57)/2=0.61

which is the golden ratio.

Project Genesis realized:
Luminosity of the sun (L) =3.826E26J/s
Mass of the sun (M) = 1.989E30Kg
The luminosity of a main sequence star like the sun is proportional to the mass by:
L is proportional to $M^{3.5}$
We introduce a constant of proportionality (a) and calculate it based on the sun, that is
$L=aM^{3.5}$
And we find $a=0.25J/(Kg)^{3.5}(s)$
Now the exponent 3.5 equals Au/Fe=196.97/55.85 where Au is gold and Fe is iron.
And the constant of proportionality, a, equals Al/Ag=26.98/107.87=0.25 where Al is aluminum and Ag is silver.
Now, Iron is the best metal for tools (early agriculture, the iron age)
And Gold is the most conductive at extreme temperatures (the space age)
Aluminum is the most abundant metal in the earth crust (electronics)
And silver is the most electrically conductive metal (let us say computers)
Thus, our project genesis equation is:
$L=(Al/Ag)M^{(Au/Fe)}$
Al/Ag in $J/(Kg^{3.5})s$
We have talked about the signifcance of the metric system where the constant of proportionality is concerned in "approaching the key".

Magnesium (Mg) is the lightest metal for construction purposes, i.e. it is good for spacecraft construction in that it provides for a low mass to thrust ratio and boron (B) is neccessary for navigation systems in the construction of system computers in that it is used to make positive type silicon, and thus the ratio of Mg/B encodes these principles in that the only terrestrial planet that can be colonized is mars from earth and the the ratio is equivalent to the ratio of the earth's escape velocity to that of mars, the velocities needed to be attained to not fall back to these respective planets. These ratio's are nearly equivalent as well to oxygen gas (O_2) to N (nitrogen) which are the components of chemical aeronomy.

$(O_2)/N=32.00/14.01=2.33\sim(v_e)/(v_m)=11.2/5.0\sim Mg/B=2.25$

where v_e=escape velocity of earth and v_m that of mars.

Mg=24.31amu and B=10.81amu

Plants are the link between humans and the sun, they convert energy from the sun into sugar, $C_6H_{12}O_6$. A gram of this, called glucose, can yield 15.6kJ of energy when burned. The reaction is:
C6H12O6+6O2--->6CO2+6H2O

(15.6E3J/g)(3E-22g/molecule)=
4.68E-18J/molecule
molar mass of glucose:

C6H12O6=6(12.01)+12(1.01)+6(16.00)=180.18amu

(180.18g/mol)(1mol/6.02E23molecules)=3E-22g

Luminosity of sun (L)=3.826E26 J/s

(365days)(24hrs)(60min)(60s)=3.15E7 s/yr

L(s/yr)=(3.826E26J/s)(3.15E7s/yr)=1.2E34 J/yr

r=earth's distance from the sun=1.5E11m

R=sun radius=7E8m

LR^2/r^2-(1.2E34J/yr)(7E8m)^2/(1.5E11m)^2=2.6E29J/yr

This last calculation is energy per year in sun light at earth orbit where a year is the time it takes for one revolution of the earth about the sun.

(2.6E29J/yr)(molecules/4.68E-18J)=5.6E46molecules/yr
(5.6E46molecules/yr)(2.99E-22g/molecule)=1.67E25g/yr

This last figure is the amount of glucose that can be produced on earth over one complete orbit of the earth around the sun (one year).

Earth mass=5.976E27g

5.976E27g/1.67E25g=357

M_j=mass of jupiter=318
M_e=mass of earth1.00

(M_j/M_e)(N/O)~357

where N=14.01, O=16.00

N is nitrogen and O is oxygen.

0.36=N/K

where N is nitrogen, the most abundant element in the earth atmophere and is key to nitrogen fixation and K is potassium, key to photosynthesis.

(surface area of earth/2)=4(pi)(6.38E6m)^2=2.6E14m^2

Neccessary in making the sugar from sun light is chlorophyll, pigements in leaves that carry out photosynthesis. In order for plants to make clorophyll, they need Potassium (K).

The way I did this calculation, may be important. For instance, when computing the number of seconds in a year, I used the 365 day year, which does not account for leap year every four years, where an extra day is added to the year.

We have shown in "genesis itself" that the glucose made by a plant can produce

4.68E-18J/molecule of energy when burned. Hydrogen gas burned in oxygen in the reaction (rocket fuel):

H_2+(1/2)O_2--->H_2O

can produce 120 kJ of energy per gram

$H_2 = 2.02$

$(2.02 g/mole)(1\ mole/6.02E23 atoms) =$

$3.35E-24 g/(molecule H_2)$

$(3.35E-24)(120E3) = 4.02E-19 J/molecule H_2$

$(4.68E-18)/4.02E-19 = 11.64 = W/O$

where W = tungsten and O = oxygen

$W = 184$ and $O = 16$

Tungsten is used to make light (i.e. is the filament of a light bulb) and oxygen is used to burn fuel, whether it be a sugar or hydrogen gas.

After crafting tools of stone, humans began to use bronze, an alloy of copper and tin. Iron was the first non alloy to be used for the same. So far we have considered carbon, oxygen and nitrogen. Here is the key to project genesis:
$(Cu+Sn)/Fe = 2(C+N)/O$ and $H \sim 1.00$
$NaCl$ = table salt = $22.99 + 35.45 = 58.44$ and $Ni = 58.69$ = nickel
Thus $NaCl = Ni$
Bronze = $Cu + Sn$ and Brass = $Cu + Zn$
Bronze/Brass = $182.26/128.94 = 1.414 = sqrt(2)$

Cu = copper, Sn = tin, Zn = zinc
$Cu = 63.55$, $Sn = 118.71$, $Zn = 65.39$

$Fe = 55.85$
With the advent of plant life, the earth began to accumulate oxygen in its atmosphere. Before that, there was a lot of CO_2. Current levels are 21% oxygen, 78% nitrogen. I suggest that the ideal levels are 25% oxygen, 75% nitrogen so that

$[(CO_2)/(O_2)][(\%N_2)/(\%N_2+\%O_2)] \sim [sqrt(2)](3/4) \sim 1.00$

We have said that:

Jupiter (meaning) reservoir/system/software=technology
Shelter: regular hexagon
Mars (meaning) software/system=magic
Food: square
Saturn (meaning) hardware=industry
Self: equilateral triangle

The dual of the regular hexagon is the equilateral triangle, the dual of the square is itself, and the dual of the equilateral triangle is the regular hexagon.

Thus, to continue, technology gives birth to industry and industry to technology and magic gives birth to magic. Thus Mars stands alone while Jupiter and Saturn are coupled.

We can now say that Mars represents godliness, Jupiter creation and Saturn rebirth.

Mars is a sort of extension of Earth – after modified—while perhaps Jupiter and Saturn are sources of hydrogen gas for rocket fuel for surface to orbit operations, necessary for building interplanetary mother ships in Mars and Earth orbit that are nuclear, preferably using the very same hydrogen for their propulsion as well. Probably after entering this phase, if we do, it will be when we will start making some breakthroughs in propulsion, hyperdrive—antigravity—that can carry us to the stars within a reasonable amount of time.
It may be that in a ceratain respect we can draw the following energy comparison:

(nuclear power)/(chemical power)=(coal)/(fire wood)=(hydro)/(wind)

(nuclear/chemical)=Jupiter
(coal/firewood)=mars
(hydro/wind)=saturn

In a sense nuclear power (H, or hydrogen) is connected to coal (C) and hydro-electric (H_2O) in the following reaction:

$C(s)+H_2O(g)$--‡$CO(g)+H_2$

Which is used to convert coal into the clean burning H_2. Keep in mind it is the carbon in coal that is an energy source, and it is hydrogen that a main sequence star begins to fuse. CO – the poisonous gas carbon monoxide—is used to separate iron from its ore.

And as sugar burns fast, carbohydrates slower but longer, and proteins slowest but longest, we have Carbohydrates correspond to Jupiter, proteins to mars, and sugars to Saturn.
I propose that the igloo represnts Saturn, the tipi Jupiter and the yurt mars. As such we have related heat and light, food and habitats to these three key planets.
Question: thus eating meat in a yurt and burning coal in it to keep you warm somewhere in Mongolia has what to do with mars?

Math Notebook of Ian Beardsley

Formulas Derived from the Parallelagram

Remarks. Squares and rectangles are parallagrams that have four sides the same length, or two sides the same length. We can determine area by measuring it either in unit triangles or unit squares. Both are fine because they both are equal sided, equal angled geometries that tessellate. With unit triangles, the areas of the regular polygons that tessellate have whole number areas. Unit squares are usually chosen to measure area.

Having chosen the unit square with which to measure area, we notice that the area of a rectangle is base times height because the rows determine the amount of columns and the columns determine the amount of rows. Thus for a rectangle we have:

$$A = bh$$

Drawing in the diagonal of a rectangle we create two right triangles, that by symmetry are congruent. Each right triangle therefore occupies half the area, and from the above formula we conclude that the area of a right triangle is one half base times height:

$$A = (1/2)bh$$

By drawing in the altitude of a triangle, we make two right triangles and applying the above formula we find that it holds for all triangles in general.

We draw a regular hexagon, or any regular polygon, and draw in all of its radii, thus breaking it up into congruent triangles. We draw in the apothem of each triangle, and using our formula for the area of triangles we find that its area is one half apothem times perimeter, where the perimeter is the sum of its sides:

A=(1/2)ap

A circle is a regular polygon with an infinite amount of infitesimal sides. If the sides of a regular polygon are increased indefinitely, the apothem becomes the radius of a circle, and the perimeter becomes the circumference of a circle. Replace a with r, the radius, and p with c, the circumference, and we have the formula for the area of a circle:

A=(1/2)rc

We define the ratio of the circumference of a circle to its diameter as pi. That is pi=C/D. Since the diameter is twice the radius, pi=C/2r. Therefore C=2(pi)r and the equation for the area of a circle becomes:

A=(pi)r^2

Math Notebook of Ian Beardsley

(More derived from the parallelogram)

Divide rectangles into four quadrants, and show that:

A. (x+a)(x+b)=(x^2)+(a+b)x+ab
B. (x+a)(x+a)=(x^2)+2ax+(a^2)
A. Gives us a way to factor quadratic expressions.
B. Gives us a way to solve quadratic equations. (Notice that the last term is the square of one half the middle coefficient.)

Remember that a square is a special case of a rectangle.

There are four interesting squares to complete.

1) The area of a rectangle is 100. The length is equal to to 5 more than the width multiplied by 3. Calculate the width and the length.
2) Solve the general expression for a quadratic equation, $a(x^2)+bx+c=0$
3) Find the golden ratio, a/b, such that $a/b=b/c$ and $a=b+c$.
4) The position of a particle is given by $x=vt+(1/2)at^2$. Find t.

Show that for a right triangle $(a^2)=(b^2)+(c^2)$ where a is the hypotenuse, b and c are legs. It can be done by inscribing a square in a square such that four right triangles are made.

Use the Pythagorean theorem to show that the equation of a circle centered at the origin is given by $r=x^2+y2$ where r is the radius of the circle and x and y the orthogonal coordinates.

Derive the equation of a straight line: $y=mx+b$ by defining the slope of the line as the change in vertical distance per change in horizontal distance.

Math Notebook of Ian Beardsley

Triangles

All polygons can be broken up into triangles. Because of that we can use triangles to determine the area of any polygon.

Theorems Branch 1
1. If in a right triangle a line is drawn parallel to the base, then the lines on both sides of the line are proportional.
2. From (1) we can prove that: If two triangles are mutually equiangular, they are similar.
3. From (2) we can prove that: If in a right triangle a perpendicular is drawn from the base to the right angle, then the two triangles on either side of the perpendicular, are similar to one another and to the whole.
4. From (3) we can prove the Pythagorean theorem.

Theorems Branch 2
1. Draw two intersecting lines and show that opposite angles are equal.

2. Draw two parallel lines with one intersecting both. Use the fact that opposite angles are equal to show that opposite interior angles are equal.

3. Inscribe a triangle in two parallel lines such that its base is part of one of the lines and the apex meets with the other. Use the fact that opposite interior angles are equal to show that the sum of the angles in a triangle are two right angles, or 180 degrees.

Theorems Branch 3

1. Any triangle can be solved given two sides and the included angle.
$c^2=a^2+b^2-2ab\cos(C)$

2. Given two angles and a side of a triangle, the other two sides can be found.
$A/\sin(a)=b/\sin(B)=c/\sin(C)$

3. Given two sides and the included angle of a triangle you can find its area, K.
$K=(1/2)bc(\sin(A))$

4. Given three sides of a triangle, the area can be found by using the formulas in (1) and (3).

Question: what do parallelograms and triangles have in common?
Answer: They can both be used to add vectors.

Math Notebook of Ian Beardsley

Trigonometry

When a line bisects another so as to form two equal angles on either side, the angles are called right angles. It is customary to divide a circle into 360 equal units called degrees, so that a right angle, one fourth of the way around a circle, is 90 degrees. The angle in radians is the intercepted arc of the circle, divided by its radius, from which we see that in the unit circle 360 degrees is 2(pi)radians, and we can relate degrees to radians as follows:

Degrees/180 degrees=Radians/pi radians

An angle is merely the measure of separation between two lines that meet at a point.

The trigonometric functions are defined as follows:

…cos x=side adjacent/hypotenuse
…sin x=side opposite/hypotenuse
…tan x=side opposite/side adjacent

…csc x=1/sin x
…sec x=1/cos x
…cot x=1/tan x

We consider the square and the triangle, and find with them we can determine the trigonometric function of some important angles.

Square (draw in the diagonal): cos 45 degrees =1/sqrt(2)=sqrt(2)/2
Equilateral triangle(draw in the altitude): cos 30 degrees=sqrt(3)/2; cos 60 degrees=1/2

Using the above formula for converting degrees to radians and vice versa:

30 degrees=(pi)/6 radians; 60 degrees=(pi)/3 radians.

Math Notebook of Ian Beardsley

The regular hexagon and pi

Tessellating equilateral triangles we find we can make a regular hexagon, which also tessellates. Making a regular hexagon like this we find two sides of an equilateral triangle make radii of the regular hexagon, and the remaining side of the equilateral triangle makes a side of the regular hexagon. All of the sides of an equilateral triangle being the same, we can conclude that the regular hexagon has its sides equal in length to its radii. If we inscribe a regular hexagon in a circle, we notice its perimeter is nearly the same as that of the circle, and its radius is the same as that of the circle. If we consider a unit regular hexagon, that is, one with side lengths of one, then its perimeter is six, and its radius is one. Its diameter is therefore two, and six divided by two is three. This is close to the value of pi, clearly, by looking at a regular hexagon inscribed in a circle.

The sum of the angles in a polygon

Draw a polygon. It need not be regular and can have any number of sides. Draw in the radii. The sum of the angles at the center is four right angles, or 360 degrees. The sum of the angles of all the triangles formed by the sides of the polygon and the radii taken together are the number sides, n, of the polygon times two right angles, or 180 degrees. The sum of the angles of the polygon are that of the triangles minus the angles at its center, or A, the sum of the angles of the polygon equals n(180 degrees)-360 degrees, or

A=180 degrees(n-2)

With a rectangular coordinate system you need only two numbers to specify a point, but with a triangular coordinate system—three axes separated by 120 degrees—you need three. However, a triangular coordinates system makes use of only 3 directions, whereas a rectangular one makes use of 4.

A rectangular coordinate system is optimal in that it can specify a point in the plane with the fewest numbers, and a triangular coordinate system is optimal in that it can specify a point in the plane with the fewest directions for its axes. The rectangular coordinate system is determined by a square, and the triangular coordinate system by an equilateral triangle. They are the basis for many mosaics in Moorish castles, such as those in the Alhambra in Spain.

From the Physics Notebook of Ian Beardsley

F=ma M=mv v=x/t

F=Force M=momentum m=mass v=velocity x=distance t=time a=acceleration

M=mv=m(x/t) a=dv/dt a(dt)=dv v=at v=dx/dt dx/dt=at dx=at(dt) x=(1/2)at^2

x=x_0+vt+(1/2)at^2

int[x^n] 0 to x =(x^(n+1))/(n-1) and (d/dx)x^n=nx^(n-1)

K=kinetic energy U=potential energy C=constant

K=(1/2)mv^2 U=mgy h=height

K+U=C mgh=mgy+(1/2)mv^2 or (1/2)m(v_0)^2=U+K where v_0=initial velocity

Work=W= and U=-W

Thus work is the distance traveled or moved by the component of the force in that direction, and potential energy is the negative of the work. Use the definition for work and the chain rule for derivatives to show that kinetic energy (energy of motion) is as given above. The chain rule is:

dv/dt=(dv/dx)(dx/dt)

A ball rolling on an incline will stay in motion until it attains the same height on another incline facing the first, even if the inclinations of the two inclines are not the same. If there is no second incline, the ball will never attain the original height and will therefore continue to roll forever, unless otherwise acted on by a force, like friction. For every force there is an equal but opposite reaction. Notice that:

$$mgh=(1/2)m(v\_0)^2$$

Let the closest star, Rigel Kentaurus, also called alpha centauri, be a metaphor for the earth, in that it is the closest star to the sun, and the third brightest in the sky and the earth is the only planet brimming with life and is the third planet from the sun. Thus we now consider the largest planet in the solar system (Jupiter) and this takes us to the brightest star in the sky, Sirius, alpha canes major, it is the fifth nearest star and Jupiter is the fifth planet. But let us associate with Jupiter as well Vega, it is the fifth brightest star. Sirius is 8.7 light years distant, and Rigel Kentaurus is 4.34 light years distant. 8.7/4.34=2.00…

If "the surface of the earth is the shore of the cosmic ocean" as Carl Sagan said, then the constellation Bootes carries its importance in the fact that it is "the boat-man". The brightest star in that constellation is Arcturus, which happens to be the fourth brightest star in our galaxy. The fourth planet is mars, and as it so happens, it is the only planet in our solar system we can colonize, in that mercury is so close to the sun that it is far too hot, venus has the same problem, but mars is the next planet after the earth, and the rest being gas giants, you can't really set foot on them. That is, we say that mercury, venus, the earth and mars are the ter-restrial planets.

First, here is a quote from the introduction of The Persian Wars, by Herodotus, by Francis R.B. Godolphin:

"A chemical formula is intelligible to anyone who knows the language; an iso-lated historical fact may correspond to a chemical symbol, but an historian's for-mula requires a great deal more than the juxtaposition of several such symbols to be intelligible."

There exists in Oregon a triad of species, as I like to call it, and they are:

Red Back Vole: by depositing its feces at the base of trees this small mammal innoculates the tree against disease allowing old growth forests to exist, becuase it eats a fungus that grows there.

Flying Squirrel: a flying mammal, it is very rare.

Northern Spotted Owl: At the top of the food chain of an old growth forest, it is an indicator species, or barometer for how the ecosystem is doing. It depends on old growth forests because it makes its nest in snags, or trees so old that the top has broken off and provides a flat surface for it to do so. This bird is endangered.

The best way to explain this is to tell it in story form. I was looking for where standard temperature and pressure (STP) occurs the most, where STP is the temperature at which water freezes (32 degrees F) and the atmopheric pressure at sea level (one atmosphere). I knew california's coast was too warm and washington state's too cold, as I have spent time in both, so I looked towards oregon. Tillamook—where they make the fabulous cheese—was two degrees too warm on the average in january, so I estimated further north, somewhere between Nehalem and Astoria on the northern oregon coast. The record high, by the way, for oregon, was 119 degrees F in Pendelton on Aug 10 1898 and the record low was minus 54 degrees on Feb 10, 1933 in Seneca. The average of these two is 32.5 degrees F, just a half of a degree over freezing.

Now at STP the molar volume of a gas is 22.4 liters and one mole is 12 grams of carbon. What then, is the molar volume in quarts? Converting, you will find it is 23.67864693 qt for there are 946 mL in a qt and 1000 mL in a liter. Notice that 23.67864693 has all of the numbers between 3 and 9 except for five, after the decimal, and all of the numbers between 2 and 9 except for 5 in the whole number. It is equal to 11200/473 and the decimal part is 321/473. Now the question I ask is what is a quart? Well it is one fourth of a gallon. Then, what is a gallon? I looked in the dictionary and it said "origin unknown". Let us consider the molar volume then, 23.6 qt, and find the element whose density in pounds per quart equals the molar volume in quarts, and we find it is lead, chemical symbol Pl for plumbus, and hence the word plumbing because the romans used lead for their plumbing. The density of lead is 11.4 g/ml, and a pound (lb) is 454 grams, and there are 946 ml in a quart.

Now in light of this a liter is one kilogram of water, at 4 degrees C, and a liter is a cube with sides one tenth of a meter, and a meter is one ten millionth of the distance from the north pole to the equator. Thus, it would seem lead and water are related in more way than one. And what are Nehalem and Astoria, these cities in oregon? They bring to mind Akkadia and the Nefilim of Nibiru, the latter being held by sumerian myth to be the people and planet responsible for the creation of our planet and ourselves respectively as translated from cuniform tablets.

Project Genesis defined: It is Assyrian in nature. Here is a quote from Carl Sagan's Cosmos.

"In ancient times, in everyday speech and custom, the most mundane happenings were connected with the grandest cosmic events. A charming example is an incantation against the worm which the Assyrians of 1000 B.C. imagined to cause toothaches. It begins with the origin of the universe and ends with a cure for the toothache."

Now the elements are in no way mundane, but as we began connecting their properties to things on a "cosmic scale" we perhaps shall soon see that everyday events are connected to those connections. We are operating on a grander scale at this point in time, than the mundane, like the constructions of pyramids, and the origins of agriculture, but soon we will hopefully understand what brushing our teeth in the morning means, in a deep sense and in relation to the structure of the universe, and as such unlock the mysteries of origin and destiny. Our advantage is that we are operating empirically as opposed to our Assyrian friends. We are doing much more than searching for a cure for the toothache, but are trying to save humanity, by opening the key to reproducing the system that keeps us alive.

We are all familiar with the pyramids in Mexico

There is teotihuacan in Mexico City and

Giza outside of Cairo, I believe

I have been told that they built pyramids in India

and that all of these lie in the so called "sun belt"

Now looking at a map, teotihaucan and Giza are both within the tropic of cancer and the first lattitude line after that, plus 30 north. I have not been able to find anything about the pyramids of India on the internet.

Now the lattitude plus 30 degrees is interesting because 30 degrees is the angle formed by drawing in the altitude of an equilateral triangle. Cairo is exactly on this lattitude, and, as I said, so are its pyramids.

Until about 8,000 B.C.

Man followed the herds as he hunted which followed the seasons

Some ten thousand years ago he began to settle in communities and farm and ranch and began to work regularly

This most likely began in the middle east, jericho being the oldest known site of a cultivating community.

Its lattitude is about the same as central California's, the most productive farming place in the world today.

And as early a 1500 B.C. in the middle east, men began to smelt iron ore. This represented a revolution in weapon and tool making. Iron is the second most abundant metal in the earth's crust comprising 4.7% of it. Heavy it holds a sharp edge. The earth's crust is mostly silicon, used to make integrated circuits, diodes, and transitors. The most abundant metal is aluminum at 7.5% of the earth's crust. As can be seen, all we need is there, and silicon is mostly made useful in silicon valley of california. There it is doped with boron or phosphorous. Computer chips are made of silicon. Thus Iron is useful for agriculture and silicon for technology. An atom of iron is twice as heavy as an atom of silicon. Silicon is 25.7% of the earth's crust.

When they speak of Atlantis, perhaps it is a methphore for Ebla which was on about the same parallel as Los Angeles, which seems to have had a parallel function, as their economy was based on the manufacture of textiles and metal works.

Of importance, I believe, is what was going on when Vega was the "pole star", Alpha Lyra, the fifth brightest star in the sky. Polaris, today's pole star is relatively faint. Vega will again be the pole star in 14,000 A.D. Vega was the pole star some 12,000 years ago, when the "land bridge" existed during the last ice age when the American Indians theoretically crossed it to the New World. It would seem to me that Vega and Polaris switch the honorable position every 13,000 years as they are in opposition in the precession of the Earth's axis, which completes a full cycle every 26,000 years. Alpha Draconis, or Thuban, the brightest star in Draco was the pole star some 4600 years ago when the Egyptians were building their pyramids.

So far we have been very accurate where the solar system is concerned, but we just don’t have accurate data to do anything on a scale as grand as the galaxies, their clusters and the dimensions of the universe, so I am left with no other alternative than to be silly in this area.

So, let me be an artist/comedian for just a minute and give the intuition free reign to draw a fuzzy picture in poem form:

The hubble constant is no doubt somewhere between 50 and 100km/sec/Mpc, but is more than likely 60 to 80km/sec/Mpc, and no doubt 70km/sec/Mpc +/- 5%. It is the expansion rate of the universe. The dinosaurs became extinct about 65 million years ago. Now let us calculate how much the universe has expanded within that amount of time at that rate.

The time: 65E6yrs

Expansion rate: 70km/sec/Mpc

There are 3.15E7sec/yr

The speed of light is: 3.0E10cm/s=3.0E5km/s

Km/ly=9.45E12

Hubble constant (H_0)=7.0E(-11)(ly)/yr/ly

Theoretical age of the universe by calculating it based on the temperature to which it has cooled (2.76 degrees K): 15 billion yrs. That makes it 15 billion ly in radius.

We have that:

$$v=D(H\_0)=15E9(ly)(7.0E(-11)ly/yr/ly=1.05(ly)/(yr)$$

Where d is distance and v is velocity, the expansion speed of the universe.

Experimental age of the universe by multiplying the speed of light by the furthest seen objects, which are 10 billion light years away: 10 billion years. That gives us:

$$V=D(H\_0)=10E9(ly)(7.0E(-11)ly/yr/ly=0.7(ly)/(yr)$$

Thus by the former the universe has expanded 1.05(65E6)=6.8E7(ly)

And by the latter: 0.7(65E6)=4.6E7(ly)

The average of these two is 5.7E7(ly)

…which is on the order of the Pegasus I cluster of galaxies in diameter.

When the dinosaurs went extinct, the Virgo cluster (core of the local super cluster of galaxies, and nearest cluster of galaxies) was just emitting the light that we are receiving today as it is 64 million light years away from us, it contains the stunning sombrero galaxy.

The distance the Bootes galaxy cluster has receded since the dinosaurs went extinct:

(39400km/s)(65E6yrs)=(1.3E(-1)(ly)/(yr))65E6yr=8.45E6(ly),

and how far the Hydra galaxy cluster has receded since the dinosaurs went extinct:

(60,600km/s)(6.5E6yrs)=(2.0E(-1)(ly)/(yr))(65E6yr=1.3E7(ly)

which is the approximate diameter of the Virgo cluster, and Cancer cluster, Leo cluster and Gemini clusters of galaxies.

Why be so interested in the correspondences of the Virgo cluster’s various parameters with the extinction of the dinosaurs? Because it would seem the dinosaurs sudden disappearance gave rise to the intelligent mammalian life we are today, and as such, with the same laws working throughout the universe, odds may be that intelligent life elsewhere would have descended from something reptilian not mammalian, as the extinction of the dinosaurs was a fluke, like cause by the impact on earth of an asteroid. Thus the Virgo cluster may correspond to the success of reptilian life where they did not have a chance to evolve here.

A.D. 100's Ptolemy offers the idea of epicycles to explain the retrograde motion of the planets.

About 310B.C.-230B.C. Aristarchus determines the distance to the moon and that the sun is very far away using parallax.

(276-195?B.C) Eratosthenes determines the circumference of the earth by the angle cast by the shadow of a stick and the absence of a shadow in a well in another location.

1473-1543 Copernicus puts forward his model of a solar system where the earth is not at the center of the universe, but that goes around the sun along with the other planets.

1564-1642 Galileo discovers four natural satellites orbiting Jupiter, and thus verifies the idea of Copernicus that the Earth is not at the center of the universe.

1571-1630 Based on the observations of Tycho Brahe, Kepler formulates the laws of planetary motion.

1642-1727 Newton makes his universal law of gravitation from which Kepler's laws can be derived. He invents differential calculus and integral calculus, the latter simultaneously with Liebnitz.

Logica System 0
There can be no so such a thing as position in infinitely extended space, for if there are only (A,B,C) then A is referenced with respect to B and B with respect C and C cannot be referenced with respect to anything independent from A or B, therefore A has no position. If there is (A,B,C...) then A is referenced by B, and B by C, and C by D, and so on forever. A is never referenced. Another way of looking at this is a circle is the collection of points equally distant from some point called the center. It follows that in infinitely extended space the center is everywhere. And space must be infinite in extension because there always has to be another side to anything, even if it is nothing.
Here is another way of looking at the Zeno paradox:
Change in position with respect to time is motion. Anything in motion cannot exist for any amount of time at any position, otherwise it has stopped. Yet we say what the position of something is after a certain amount of time given its speed. Thus motion is impossible, even though we see it happen. But that is O.K. because we have shown there can be no such a thing as position.
Logical System 1
Prove: God is artificial.
Postulate 1: For ourselves, the planets, the stars (the galaxies are made of stars), and light to be here, something had to come into existence from nothing,

uncaused, yet that is impossible. The only other alternative is that something has always been here, and that does not make any sense. In this sense nothing can exist.

Axiom 1: It cannot be proved that what we see, hear, smell, feel or taste exists.

Definition 1: God – the source of existence.

Theorem 1: Man made God (by postulate 1, axiom 1 and, definition 1).

Definition 2: That which is man made is artificial.

Theorem 2: God is artificial (by definition 2 and theorem 1). QED.

Here is a clever argument I heard on the uk.philosophy.atheism newsgroup. I will put it into a Euclidean argument form.

Logica System 2

Prove: Nature is God.

Axiom 1: Objects require their opposite to be defined. i.e. matter is defined in terms of space: it is that which occupies space and has mass.

Theorem 1: The set of everything is beyond explanation because there exists nothing outside of it to define it (by axiom 1).

Definition 1: God—that which is beyond explanation.

Definition 2: Nature—everything.

Theorem 2: Nature is beyond explanation (by theorem 1 and definition 2).

Theorem 3: Nature is God (by theorem 2 and definition 1). QED.

Hypothesis: For every thesis the antithesis exists, even if neither make sense. In other words, to quote Nielhs Bohr: “The great truth is a statement whose opposite is also true.” (By induction contrasting the last theorems of logica system 1, and logica system 2).

The truths that we accept, are, more than anything, dependent on our definitions. Different cultures represent different logic systems. Logica System 0, Logica System 1, and Logica System 2 represent different hypothetical cultures. Notice how one and two come to a seemingly opposite truth. Logica System 0 we will call universal: it’s truths are accepted by all cultures, let us say.

The sun radiates 1.2E34J/yr and

plancks constant is 6.626E-34(J)(s) (minimum energy packet)

(1.2EJ/yr)(6.626E-34J*s)/(yr/3.15E7s)=2.5E-7J^2

(1.01g H/mol)(mol H)/(6.02E23 atoms H)=1.68E-24gH/atom

rest energy of a hydrogen atom:

(1.68E-24g)(3.0E8m/s)^2(kg/1000g)=1.5E-10J

(2.5E-7J^2)/1.5E-10J=1.67E3J

The specific heat of water is 1.00cal/g*deg C and

1.67E3J=3.99E2cal

the normal human body temperature is 37 deg C

(3.99E2cal)(deg C)(g)/(1.00cal)(37deg C)=10.78gH_2O

10.78g=1.078E-2Kg

(1.078E-2)(15807m)(9.8m/s^2)=1.67E3J

15807m=16km

which is right in the middle of that narrow layer of atmophere between the tropo-sphere and stratosphere, called the tropopause.

Now 16km is 19 kHz, or very low frequency (VLF) which corresponds to wave-lengths between 100km-10km. It is a radio frequency and happens to be the fre-quency at which radio waves can penetrate water 10-40 meters deep (10-20kHz) and thereby is used by submarines near surface for communication. Now this was based on the speed of light in a vacuum, but in anycase we have connected human activity to natural law, physical and biological, and are thus bordering on psychohistory.

Now the normal human body temperature is about the difference between the coldest day and warmest day of the year in Red Bluff, California, the agricultural hub of America, if not the world. That is the average temperatures swing between about freezing (of water), 0 degrees C and 37 degrees hot throughout the course of a year. In fact between 1961 and 1990 the average yearly normalized temperature was 17.1 degrees C. (37 deg C - 0 deg C)/2=37/2=18.5degC, an accuracy within 92%. Red Bluff is in Tehama County, which is at about 40degN, 122degW. (See www.worldclimate.com)

We apply the triangulation of Chris Darrow, where three relevant locations are chosen on a map, are connected and the associated power spots pertaining to the corners are revealed. Bear in mind that a triangle is the minimal structure that encloses an area.

I have chosen Nehalem, Oregon as it is where I estimate that standard temperature and pressure occurs most frequently, that is freezing temperature of water and 1 atmosphere of pressure. I have chosen as a second point Red Bluff, California as its annual average temperature is the mean of freezing and the normal body temperature, 37 deg C. I have chosen as the third point Bend, Oregon, as it is "land's end" in the high desert the last city before the vast expanse to Idaho, nearly. Connecting these points we have a triangle who's "center of gravity" seems to be the Cascade Mountain range Summit at Crater Lake: http://www.crescentlakeresort.com/

Earth-moon distance: 3.84E10cm=R
Earth-sun distance: 1.496E13cm=r
Earth radius: 6.38E8cm=E_r
Moon radius: 1.738E8cm=M_r

Apparent diameter moon=apparent diameter sun=0.5 deg

(360)/(0.5) =720 moon diameters/celestial equator = 720 sun diameters/celestial equator

$(E_r)/(M_r)$=11/3 r/R=4675/12

(4675/12)(11/3)/720=2

$(E_r/M_r)(r/R)(D/C)=(O_2)/(CH_4)=(r_u)/(r_s)$

C=degrees in circumference of a circle
D=apparent diameter of moon=apparent diameter of sun in degrees.

…r_u=mean distance from sun of Uranus
…r_s=mean distance from sun of Saturn

We have essentially shown that:

$$((E\_r)/(M\_r))((1\ deg\ in\ rad))=(8(pi\ rad))(R/r)$$

Where, again, E_r=earth radius, M_r=moon radius
C=radians in circumference of a circle, R=earth-moon distance
r=earth-sun distance. The last means that
$$((E\_r)(r))/((M\_r)(R)(360))=4$$
Now 360 are the degrees in a circle and
4 weeks is a complete revolution of the moon about the earth more or less on the average (time beteen new moons). 360 degrees are convenient for the units inwhich to divide a circle, because of its divisible properties. (i.e. it is divisible by 120, 60, 45, 30, and 90, which are the angles in special triangles.)

The specific heat of water, the energy required to raise the temperature of a gram of it one degree centigrade is (1.00 calories/gram-degree centigrade).

The luminosity of the sun is 3.826E26J/s.

(365days)(24hr./day)(60min/hr.)(60s/min)=3.15E7 seconds per year.

The annual output of energy by the sun in light, that is, over one complete revolution of the earth about the sun is:

(3.826E26J/s)(3.15E7s)=1.2E34J/yr

One calorie (1 cal)=4.184 Joules (J)

(1.2E34J)(cal/4.184J)=2.87E33cal

That is the annual solar output of energy in light in calories.

Water freezes at zero degrees C at 1 atmosphere of pressure, or, at sea level in other words, and the normal human body temperature is 37 degrees C. Thus using our specific heat of water as a unit factor, we have, since 37C-0C=37C:

$$(2.87E33cal)(1/37C)(g\text{-}C)/cal)=(7.76E31g)(Kg/1000g)=7.76E28Kg$$

A so-called enthalpy calculation will show that a gram of sugar as made by plants (C6H12O6) will produce 1.56E4J of energy when burned in oxygen. Thus,

$$(7.76E28Kg)(1.56E4J/g)(1000g/Kg)=1.2E36J$$

Thus if over the complete revolution of the earth about the sun, the total energy output of the sun in light, raises water from freezing to the normal human body temperature at the freezing of water and sea level, then that would correspond to 7.76E28Kg of water, which in mass of photosynthetically produced sugar, burned, is 1.2E36J of energy.

Now Jupiter has a mass of 1.9E27Kg, and a mean orbital velocity of 13060m/s. This means its kinetic energy due to its orbital motion, or energy of motion, as given by $(1/2)mv^2$ is:

1.62E35J

$$(1.2E36J)/(1.62E35J)=7.41$$

$$7.41\sim Au/Al=196.97/26.98=7.3$$

Where Au is the molar mass of gold, and Al is the molar mass of aluminum.

Gold is special for its unique electrical properties and aluminum is the most abundant metal in the earth crust.

Absolute zero, or as cold as it gets in other words, is -273 degrees C, and, the normal human body temperature is, as we said, 37 degrees C. That makes us 8.378 times warmer than absolute zero, which corresponds to the ratio of platinum to sodium: 195.08/22.99=8.4854=Pt/Na. I believe this relationship to be the key to maintaining a habitable habitat (temperature) for humans inside, and perhaps even outside the ship. (Consider platinum and sodium).

“heart of genesis continued” leaves us asking what mars might have to do with the earth. Let us calculate its kinetic energy:

$(1/2)mv^2=1.868E29J$

because:m=6.418E23kg and v=24.13km/s

Let us compare this to the energy required to move the moon from earth to its present orbit:

$F=-(GmM)/r^2$

$W=-(GmM)int\{0$ to $r\}r^-2dr=GmM/r$

$G=(6.672E-8dyn*cm^2/g^2)$ M=2.35E25g   m=(5.976E27g)   r=3.84E10cm
$J=10^7ergs$

I get:7.63E28J

Equatorial radius of the moon equals: 1.738E6m

Equatorial radius of mars equals: 3.393E6m

1.868E29/3.393E6m=5.5E22newtons

7.63E28J/x=5.5E22newtons

x=1.387E6m=1,387km=1.387E8cm

The natural satellite of Jupiter, that which has the best chance of having life for a natural satellite, is Europa. It has a radius of 1,565km.

(1,387km/1,565km)100=88.6%

(1,565km/1,738km)100=90%

You really have to think about this one to get the meaning, and I have to admit, at the moment I am quite lost, but can see how to interpret it and will do so time permitting. It is a matter of going through it step by step.

The temperatures of Mongolia alternate between extremes of 57 degrees F and 96 degrees F, a yearly average of 76.5. Unique for its latitude. Eugene Oregon, for example, at the same latitude has a yearly average of 53 degrees F varying from 36 in the winter to 70 in the summer (on the average). This average yearly temperature in Mongolia is the same as Ecuador at the equator, 0 degrees latitude, almost always 75 degrees in the low lands, yet the location of Mongolia is half way up the Earth from the equator (45 degrees north). Alaska as far north as you can go on the American continent has a yearly average of 18 degrees farenheit. Europa, at Jupiter, is about 39% the way down in temperature varying from the sun to the last planet Pluto. It is 25 times colder at Jupiter than it is at Earth, by the inverse square law. Thus we see tha Europa is the Mongolia of the solar system, as it may be, below its icy surface, their is liquid water and possibility of life, condintions unlike anywhere else in the solar system. As it turns out, Mongolia mines coal.

A convenient way of looking at this is:

Kinetic energy of Mars to the radius of Mars equals the potential energy of the moon to the radius of Europa.

I should note that this potential is that if the acceleration were the same at lunar orbit and earth surface. Here we will account for the change in acceleration from earth to lunar orbit. Earth radius is 6.3785E8cm. Thus we must subtract 7.63E28J from 1.468980E30J, and we get:1.39268E30J. Let's just call it 1.39E30J. We have simply made the same calculation as above for the so called "potential energy", but with the earth radius for r, and subtracted from it our original value.

Comparing this to the kinetic energy of mars (due to its orbital motion) we have:

1.39E30J/1.868E29J=7.44

7.44~Hg/Al=200.6/26.9=7.45

Hg= atomic weight of mercury and Al that of aluminum.

The Earth atmosphere is nearly 25% oxygen and 75% nitrogen. Oxygen nearly weighs the same as nitrogen. Both occur as diatomic molecules. O=16.00 and N=14.01~14.00.
Let us say imagine that there was such a thing as an air atom, then, its relative mass would be:

$(16.00)(0.25)+(14.00)(0.75)=4+10.5=14.5=air.$

If it was diatomic, then air2=29.00: somewhere between silicon and phosphorus. Silicon is used in combination with phosphorus for making diodes, and ultimately transitors and integrated circuits.

There are 6.02E23 atoms/mole

$(14.5.00g/mole)(mole/6.02E23atoms)=2.4E-23g/atom$

Or,

4.8E-23g/molecule

Concept Habitat by Ian Beardsley: optimal shelter design and ecosystem/ecology dynamics—nature in harmony with life. As the drawing below details, building bases are one of the three regular polygons that tessellate, allowing for a travel grid between structures. The four-faced pyramid is the most stable, the cube with octet truss to direct snow and rain to the ground, is the easiest to build, and the regular hexagon translated vertically uses the least materials. The idea is to build upward as much as possible to maximize available land. At the crux of a viable ecological plan, is to ensure that the four natural cycles: nitrogen, phosphorus, oxygen/carbon and water, maintain their renewable aspect. Low entropy means getting the most for the energy spent, which is important in a closed system (i.e. in a closed system energy goes from usable to unusable, and to make it usable again, requires energy from outside the system.) One can see the importance if the universe is finite, or if other systems are difficult to impossible to access. Currently we are trapped in the earth-sun system, more or less. Scroll down to next page for pictures.

>
> Archimedean tessellations use two or more different regular polygons to tile the plane.
> As such we can introduce the regular pentagon in combination with the
> equilateral triangle among other shapes. If we do not restrict ourselves to
> equal sided equal angled figures, the possibilities become infinite.

Aside from regular tessellations, and Archimedean tessellations, we have another family of shelters, which consist of conical (tipi) spherical (igloo) and
cylindrical (yurt). Thus there are three BASIC families of habitats. I have drawn three cities, each of these families, in the photo section below.

The next four pages are what I call the basic habitat families: an example of some elements of the Archimedean family, an example of the elements of the curved family (an igloo, a yurt and a tipi.)An explanatory drawing of some dynamics of the regular tessellation family, and a regular tessellation family city.

Notice that the curved family habitats are nomadic while the regular tessellations family are European. What I call the Archimedian family are only how we could imagine Atlantis would have been had it existed. I have added the Hogan and Zome, the latter modeled after a pinecone.

◆   ◆   ◆

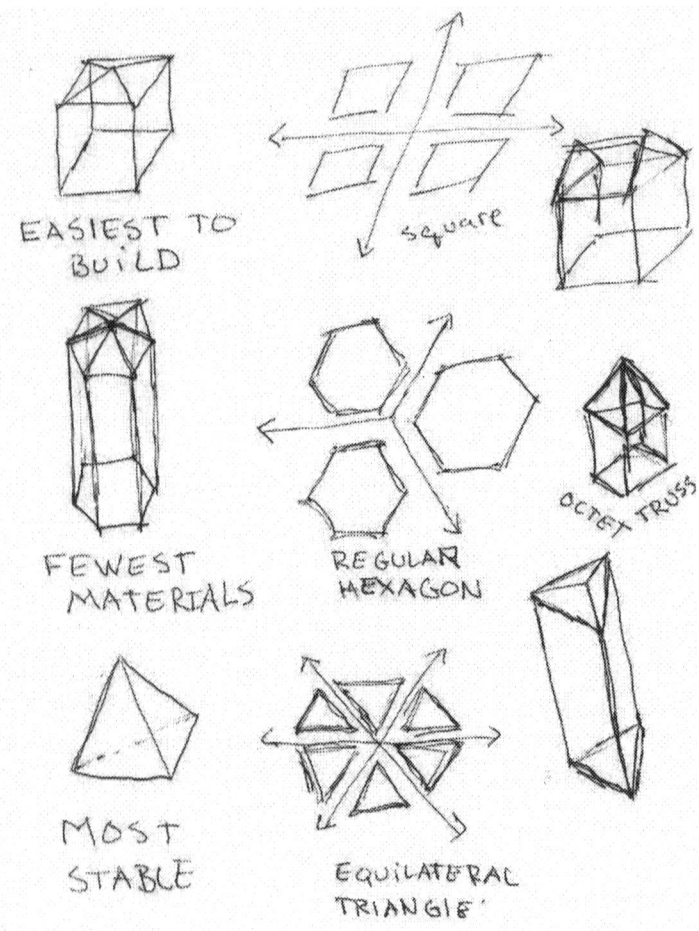

EASIEST TO
BUILD

square

FEWEST
MATERIALS

REGULAR
HEXAGON

OCTET TRUSS

MOST
STABLE

EQUILATERAL
TRIANGLE

City Built
of
Regular Tessellators

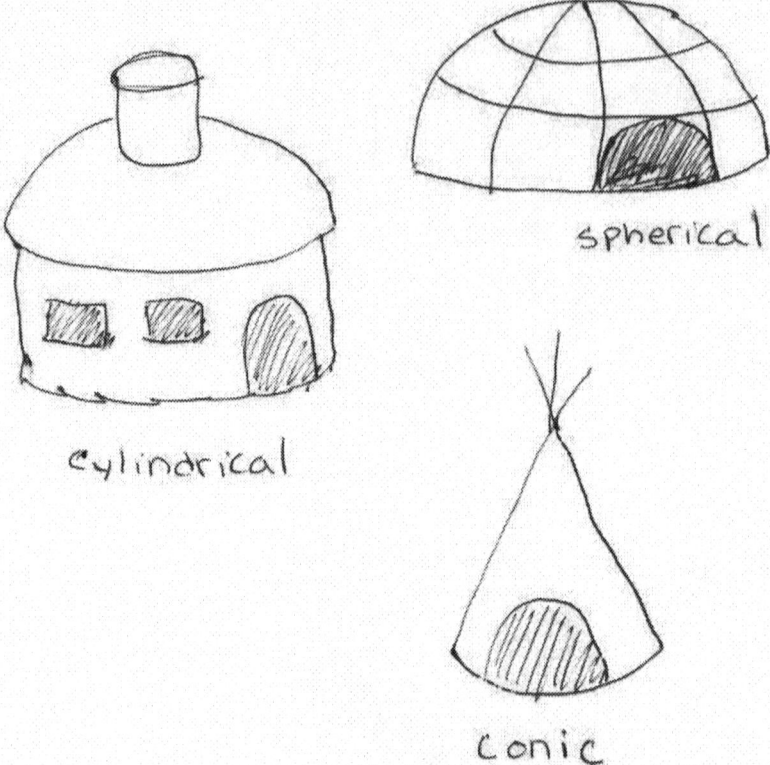

spherical

cylindrical

conic

1 - uniform family

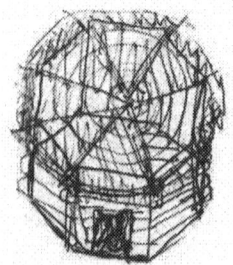

HOGAN
(OCTAGON WITH conic roof)

spiral Family

modeled
after
Pincone

Zome

date	time	moon set	moon rise	set ratio	incline set	incline rise
9/24/04	1:20AM	233 deg W		0.647	20 degrees	
9/26/04	3:10AM	231 deg W		0.641	22 degrees	
9/27/04	4:45AM	250 deg W		0.694	4 degrees	
9/27/04	7:35PM		111 deg SE			9 degrees
9/28/04	8:10PM		96 deg SE			14 degrees
9/28/04	8:17PM		93 deg SE			20 degrees
9/29/04	8:40PM		87 deg NE			17 degrees
9/30/04	8:58PM		82 deg NE			9 degrees

"Incline set" is degrees of moon above horizon when I measured it setting.
"Incline rise" is degrees of moon above horizon when I measured it rising.
The "set ratio" is the ratio of degrees west the moon set compared to 360 degrees.
I hypothesize the moon will set at the golden angle when it sets its furthest south
(222 degrees west, which is a "set ratio" of 0.618, the golden ratio.)

Position as measured with a compass is different than position as measured
with a telescope. A telescope measures "right ascension" which are projections of
longitude lines onto the celestial sphere, whereas a compass measures along the
horizon in an apparent plane clockwise starting at north. I have also made a sextant for measuring inclination. This is different as well from the astronomer's
declination, which is measured by projections of latitude lines onto the celestial
sphere. Thus these are not ephemeral positions.

Now it is entirely possible that ancient cultures might have noticed times
when the moon set at an angle of rotation clockwise around the horizon from
north that is the same angle of rotation around the stem of a plant by consecutive
leaves, and identified this as approximately 2/3, an approximation to the golden
angle. We shall see just when the moon sets at the golden angle this year, if it has
not already. And it is looking like it will do it this year.

The moon should measure the same setting and rising positions from any different positions on the Earth at the same time because the angle made with any
two positions on the Earth, and the moon, will be very small, because the moon

is so far away compared to any change in position on the Earth. Therefore these measurements are position independent, within the accuracy of my compass and sextant, which are both plus or minus one degree. See the picture below.

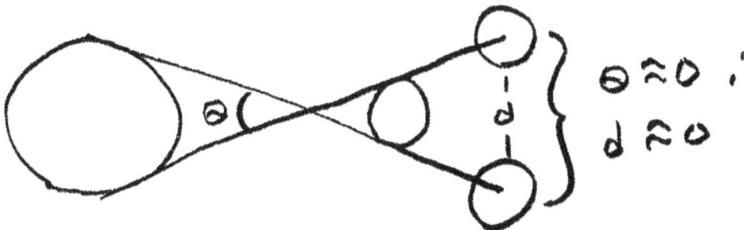

This has been "Moonset" by Ian Beardsley (unlocking the mystery of the valley.)

"Moonset" by Ian Beardsley Continued…

I used an accurate military compass with viewing slit and cross hair to make the measurements. There is a built in magnifying glass to read position. The compass position parameters above are normalized to account for the difference between true north and magnetic north, which varies with latitude and longitude. At my latitude of 34 degrees north, 117 degrees west, the magnetic declination, or degrees east that magnetic north is of true north is 13 degrees. Therefore 13 degrees were added to all of my compass readings. I made my sextant from a protractor, drilling a hole in the middle of the straight edge part from which to hang a pointer that remains pointing at the Earth's center by gravity as the readings rotate through it.

It is a curious thing about the moon that, as seen from the Earth, it appears to be the same size as the sun, nearly enough that when it passes between ourselves and the sun there is a near perfect eclipse, which allows us to observe the outer thin atmosphere of the sun, that is otherwise impossible to observe, because of the brightness of the sun's main body. When we take into account that the most spectacular meteor shower of the year, the perseids, is heralded by the heliacal rising of the brightest star in the sky, Sirius, here in Southern California on August 12, then things become quite interesting. Here in the mountains that souround this valley, the San Gabriel Mountains, occurs a rare deep blue gem stone, Lapis Azuli, which is protected by law. The only other place in the world where it occurs, are the mountains of Afghanistan.

moon	
mass	7.35E25g
radius	1,738km
density	3.34g/mL
gravity	0.166g
day	27.322d
inclination	6.68 deg

Around this house alone over the past couple of months I have been approached by three immature red tail hawks that landed in the pine tree next to me, by one black rabbit, and by a family of raccoons several evenings. I have seen a coyote in the foothills, and there are plenty of squirrels all over town. Western scrub jays and crows are highly abundant. In the mountains I have seen herds of big horn sheep.

Native to this valley, and throughout the state, is the golden poppy wild-flower, which does well in dry, harsh climates. It even flourishes in great abundance in the Mojave Desert on the other side of the mountains. There is a golden poppy reserve there, in Antelope Valley.

On 10/22/04 I measured the moon to set at the golden ratio, which happened to coincide with the village venture in Claremont, California, a sort of celebration of the fall equinox. This could be project genesis related in a connect five sense.

Arch

0-595-34490-9